Marcel Ricardo da Silva
Andréa Meiado Chiarioni

Bioinvasive Species in Environmental Education

AF757287

Marcel Ricardo da Silva
Andréa Meiado Chiarioni

Bioinvasive Species in Environmental Education

A Case Study of the Bioinvasion of Golden Mussels in the Araçatuba Region, São Paulo State

ScienciaScripts

Imprint

Any brand names and product names mentioned in this book are subject to trademark, brand or patent protection and are trademarks or registered trademarks of their respective holders. The use of brand names, product names, common names, trade names, product descriptions etc. even without a particular marking in this work is in no way to be construed to mean that such names may be regarded as unrestricted in respect of trademark and brand protection legislation and could thus be used by anyone.

Cover image: www.ingimage.com

This book is a translation from the original published under ISBN 978-620-2-03234-6.

Publisher:
Sciencia Scripts
is a trademark of
Dodo Books Indian Ocean Ltd. and OmniScriptum S.R.L publishing group

120 High Road, East Finchley, London, N2 9ED, United Kingdom
Str. Armeneasca 28/1, office 1, Chisinau MD-2012, Republic of Moldova, Europe
Managing Directors: Ieva Konstantinova, Victoria Ursu
info@omniscriptum.com

Printed at: see last page
ISBN: 978-620-6-17226-0

Copyright © Marcel Ricardo da Silva, Andréa Meiado Chiarioni
Copyright © 2025 Dodo Books Indian Ocean Ltd. and OmniScriptum S.R.L publishing group

SUMMARY

MARCEL RICARDO DA SILVA [1]

ANDRÉA MEIADO CHIARIONI

ABSTRACT: This work aims to report through a lesson on exotic species, the case study on the bioinvasion of the golden mussel, *Limnoperna fortunei* (Dunker, 1857), in the region of Araçatuba/SP between the years 2010 to 2015, as a complementary activity proposed in the CEEJA (State Centres for Youth and Adult Education) material, (Unit 1 - Interdependence of Living Beings), for high school students (1st Term) in the EJA (Youth and Adult Education) modality of the Penitentiary Progression Centre (CPP) in the city of Valparaíso/SP. Environmental education brings an understanding of the values of the environment in today's contemporary society, providing citizens with subsidies to act incisively on issues related to nature and interacting with their community. The intentional or accidental introduction of bioinvasive species into new environments leads to an imbalance in fauna and flora, with consequences for the environment, including a reduction in local biodiversity. Specifically in South America, the bioinvasion of the golden mussel through ballast water has been causing environmental and economic damage and is a cause for concern for the competent bodies in relation to its presence in the country. By relating the example of the bioinvasion of the golden mussel in the Araçatuba/SP region, from its introduction, biological characteristics and ways of preventing its proliferation to other environments, the students were able to gain a comprehensive understanding of the subject of exotic species in a contextualised and dynamic way, using environmental education teaching methodologies.

KEY WORDS: Environmental education, bioinvasive species, golden mussel.

[1] Specialised in Environmental Education at the São Luís Faculty of Education. Graduated in Chemistry. Graduated in Biofuels Technology. Author's *e-mail*: marcelricardo@professor.educacao.sp.gov.br

CHAPTER 1

INTRODUCTION

The interrelationships between the natural environment and social aspects are on the agenda in various areas of knowledge. Environmental education professionals and scholars seek to make people aware of the importance of debating the exploitation of natural resources, consumerism and the preservation and maintenance of species (COSTA; IGNÁCIO, 2014).

Environmental education is included in academic subjects to involve the whole community in the global reality of the relationship between individuals and the natural environment. It establishes the discussion of problems and the origin of environmental causes in order to raise awareness and actions among individuals towards environmental causes. The development of environmental awareness in people promotes an attitude of behaviour aimed at overcoming this reality, both in its natural and social aspects (AMBIENTE BRASIL, 2008).

A bioinvasive species is any species that is found outside its natural distribution area. Essentially, they threaten ecosystems, *habitats* and other species. Due to their competitive advantages and the absence of natural enemies, bioinvasors are able to proliferate and invade ecosystems, whether natural or anthropised (MMA, 2010).

The golden mussel *Limnoperna fortunei* (Dunker, 1857) is a bioinvasive species from China. It was introduced to South America through ballast water, where cargo ships accidentally brought larvae of this specimen into South American waters. Among the impacts caused by the golden mussel are the effects on competition for nutrients with other native filter-feeding animals, a decrease in phytoplankton biomass and turbidity, which lead to an increase in aquatic macrophytes (VON RUCKERT et *al.,* 2004).

1.1. General objective

- Present the aspects of environmental education in the global context and in Brazil in relation to the progress and implementation of its guidelines.

- To collect data on the bioinvasion of the golden mussel since its introduction to South America and bibliographical surveys that have been carried out on the subject between 2010 and 2015 in the Araçatuba region in the state of São Paulo.

1.2. Specific objective

- To use the data collected on the bioinvasion of the golden mussel in the region of Araçatuba/SP and, through an informative poster, to promote environmental education and awareness among the students of the Youth and Adult Education (EJA), of the 1st ° (Term) of high school at the Penitentiary Progression Centre (CPP) of Valparaíso/SP.

1.3. Description of practice

The educational practice was carried out at the Penitentiary Progression Centre (CPP) located in the city of Valparaíso in the state of São Paulo with the support of the link school "Escola Estadual Vicente Barbosa".

The Araçatuba Directorate of Education, through the team of Pedagogical Centre Coordinating Teachers (PCNP), provided all the encouragement to carry out the classes using pedagogical tools, especially in the technical guidance (O.T) with the mediation of the PCNP's Dorislei Aparecida Teixeira de Carvalho (Sciences), Airton Miranda Pinto Júnior (Biology) and Ângela Maria de Oliveira Barbosa Nunes (Chemistry).

The work teams involved were Professor Mareei Ricardo da Silva (Chemistry/Science/Biology), who accompanied and mediated the lecture using an illustrative teaching tool (poster), with the guidance of Master Professor Mário Marcos Lopes. Professor Sandra Maria de Melo and Professor Andréa Meiado Chiarioni collaborated on the field research and bibliographic surveys. The pedagogical coordinators Sônia Maria Lacerda and Angela Cristina Amorim da Silva and the Principal Creuza de Fátima Ervolino of the

"Escola Estadual Vicente Barbosa" school and other members of the management team who helped to provide support and guidance for the learning process.

The support and structure for the learning situation at the Teaching Unit was provided by the Unit Director Marcelo Scatena Franco of the Penitentiary Progression Centre (CPP) in Valparaíso/SP. Permission was granted by the Director of Education and Work, Pedro Valmir Berssane, the Director of Security Antônio Donizete Ferreira, the Coordinator of Education and Work Élida Regina Tegon, with the collaboration of staff members Gervano Lisboa and Mariane Tegon.

The other teachers at the "Escola Estadual Vicente Barbosa" (Vicente Barbosa State School) were fundamental in providing support in their respective subjects, providing subsidies to complement the students' studies in understanding the lesson.

CHAPTER 2

THEORETICAL BACKGROUND

2.1. History of environmental education in the world

Literature on themes related to environmental issues such as "Man and Nature", or "Physical Geography Modified by the Action of Man", both written by Georges Perkins Marsh, refers to the beginning of environmental education as a milestone in 1864. The terminology of the word "ecology" was proposed by Ernst Haeckel to define the studies to be carried out on the relationships between species and their environment (ARAÚJO, 2007).

With the advent of the "Study of the Protection of Nature in the World", organised by the International Union for Conservation of Nature - IUCN, which had been created as a result of the *Fontainbleau* International Conference in France in 1948 to debate environmental issues and set goals for nature conservation studies (ARAÚJO, 2007).

In the 20th century, the social conquests and quests initiated by the movements of minority struggles that began in 1970 were an important milestone in leveraging and putting environmental issues on the agenda. The publication of the book "Silent Spring" (1962) by the American Raquel Carson had consequences for the world's understanding of man's exploitation of nature using toxic products. As a result, international events addressed the issue of preservation and the introduction of environmental education worldwide (GALLI, 2007).

The Stockholm Conference in 1972, organised by the United Nations on environmental issues, involved 113 countries. The main issues raised at the event were denunciations of the devastation of natural resources, questioning the demographic growth of humans in relation to the occupation of the environment (SENADO EM DISCUSSÃO, 2014).

The debates that took place at the Congress involved questions about the introduction of environmental education to combat the environmental problems that existed at the time. As a result of these questions, two important environmental documents were drawn up: the "Declaration on the Human Environment" and the "World Plan of Action" (ARAÚJO, 2007).

Three years after the Stockholm Conference, Belgrade (now the capital of Serbia) held a review of environmental actions in 1975, bringing together researchers and scientists from 65 countries. The drafting of the "Belgrade Charter", which explored issues such as the eradication of poverty, the end of illiteracy, hunger, pollution, exploitation and all forms of human domination, gave rise to a new ethic on social issues. Environmental issues are associated with the international principles and guidelines of environmental education to delimit humanity's actions regarding the mass exploitation of the environment. Continuous and multidisciplinary in nature, environmental education must take into account regional differences and national interests (QUINTINO, 2013).

International discussions about the need to set up contemporary Brazilian environmental education are moving along at a slow pace.

National environmental protection legislation is a reflection of the global circumstances experienced over the last few decades, in which the nature-destroying development model generates a risk society, in the broader context of the capitalist technological society. Naturally, environmental education legislation also reflects this global concern and stands out as a mechanism for sustainable development (ARAÚJO, 2007).

2.2. History of environmental education in Brazil

The history of environmental education in Brazil was influenced by the premises of the Stockholm Conference and due to environmental concerns, the Brazilian government established the Special Secretariat for the Environment with the aim of implementing integrated environmental management. With an outdated structure, this body was unable to debate and propose new measures for the benefit of the environment (JANKE, 2012).

According to Quintino (2013) *apud* Perini (1998), training educators was on the agenda of the Conference's action plan. New methodologies were related to educating with environmental ethics intertwined, in line with what had been established on a global level. Based on this new stance that environmental educators had in establishing their teachings, three more international conferences on environmental education were held between the 1970s and 1980s.

From the 1990s onwards, the precepts of environmental education in Brazil matured. The rise began with the Tbilisi Conference. Organised by UNESCO in 1977 in the capital of Georgia, the former Soviet Union, it resulted in a final document that is the basis for the modern vision of environmental education (MMA, 2009).

In summary, the reorganisation to institutionalise the Environmental Education process takes place as follows (PRS, 2013; CADERNOS SECAD1; PCN, 1997):

- 1973: creation of the Special Secretariat for the Environment (SEMA), linked to the Presidency of the Republic.

- 1981: Institutionalisation of Environmental Education, with the National Environmental Policy (PNMA), which established, in the legislative sphere, the need to include Environmental Education at all levels of education, including community education, with the aim of enabling it to actively participate in the defence of the environment.

- 1992: The Brazilian Charter for Environmental Education was established during the Conference known as Rio 92, with the participation of the Ministry of Education and Culture (MEC). It recognised that environmental education is one of the most important tools for achieving sustainability as a strategy for the survival of the planet and, consequently, for improving the quality of human life.

- 1997: The National Curriculum Parameters (PCN) were approved by the National Education Council. The PCN are a subsidy to support schools in drawing up their educational project, inserting procedures, attitudes and values into school life, as well as the need to deal with some urgent social issues of national scope, known as cross-cutting themes: the environment, ethics, cultural plurality, sexual orientation, work and consumption, with the possibility of schools and/or communities choosing others of relevant importance to their reality (BRASIL, 1997).

- 2004: A ministerial change, the consequent creation of the Secretariat for Continuing Education, Literacy and Diversity (Secad) and the transfer of the CGEA to this secretariat, allowed environmental education to become more deeply rooted in the MEC and in the state and municipal education networks, becoming integrated with the areas of Diversity, Indigenous School Education and Education in the Countryside, thus giving environmental education greater visibility and highlighting its cross-cutting vocation.

- 2004/2007: Brazil is in line with Latin American and Caribbean countries in making international commitments to implement the Latin American and Caribbean Environmental Education Programme (PlacealO) and the Andean-Amazonian Plan for Environmental Communication and Education (Panacea), which include the countries' Ministries of the Environment and Education.

- 2013: Environmental Education in Brazil was consolidated through the National Curriculum Parameters (PCNs), where it is presented as one of the transversal themes, as can be seen in the publication Inclusão de Ciências no Saeb: documento básico (INEP, 2013), and should be worked on in such a way as to emphasise social, political, economic and ecological aspects. Thus, as a cross-cutting theme, Environmental Education should be present in all subjects, permeating their content in order to build a more integrated vision and promote understanding of socio-environmental issues as a whole.

2.3. Bioinvasive species

The process of expansion and globalisation, coupled with the intensification and speed of human movement, has contributed greatly to the breaking down of ecological barriers and consequently the significant increase in the introduction of exotic bioinvasive species into the environment (MACHADO, OLIVEIRA, 2009).

The problem of the spread of exotic species leads to homogenised environments. In this way, environments lose their unique characteristics, which local biodiversity provides, and essential ecological properties are altered. These changes are exemplified by modifications to water and nutrient cycles, productivity and the trophic chain (MMA, 2011).

Exotic species capable of overcoming barriers to colonisation, reproduction and dispersal imposed by foreign environments can become invasive species, causing impacts on invaded environments, their native species and/or human activities. This process, known as biological invasion, stems from the competitive advantage and dominance of the invading species over local species (VALÉRY *et al.,* 2008).

The survey on invasive alien species records more than 120,000 species of plants, animals and microorganisms already registered in six countries: South Africa, Australia, Brazil, the United States, India and the United Kingdom (AMBIENTE SÂO PAULO, 2011).

In Europe since 2002, in order to deal with the issue of bioinvasive species, they have established targets described in the project called: *Daisie (Deliveríng Alien Species Inventories for Europe).* The studies identified 10,992 species that have invaded Europe and with the data in hand, strategies for the prevention, management, control and eradication of invasive species have been developed (SISTEMA AMBIENTAL PAULISTA, 2009).

The distribution of these species in Brazil, according to the Ministry of the Environment, after a survey of data related to the distribution and installed capacity of the species resulted in the "National Report on Invasive Alien Species". This work identified 543 species of organisms that affect different ecosystems, including production systems and human health (MMA 2011).

In the state of São Paulo, some bioinvasive species have already been reported due to the magnitude of the impact they have on the environment (SISTEMA AMBIENTAL PAULISTA, 2009):

- European hare (Lepus *europaeus):* much larger than the native species (Sy/v/7agus *brasi/iensis* - tapiti), it competes with it and causes damage to agriculture. The occurrences reported in the state are concentrated in the south-west. The species arrived in Brazil from Argentina, where it was introduced for breeding.

- African snail *(Achiatina fu/ica):* it was introduced into the country for gastronomic reasons, and as it was unsuccessful, it was left in the environment. It consumes a large amount of vegetation, reducing its availability to native fauna and causing damage to horticulture. It is also a vector for human diseases.

- Wild boar (Sus *scrofa):* populations of this species compete with native wild pigs, alter the constitution of the soil, especially wetlands and riverbanks, cause damage to agriculture and transmit various diseases, including foot-and-mouth disease and leptospirosis. It was introduced for captive breeding, which was permitted until 1998.

- Herbaceous plant species that forms dense clumps in marshes, streams and wetlands, preventing the establishment of native species.

- Aquatic brachiaria *(Brad^aria subquadriaray* grass introduced for forage purposes in wetlands or flooded areas, but has not had the expected success as a species of economic value. But due to its high competitive capacity with native grasses, it has occupied extensive areas, causing damage to reservoirs and water bodies.

- The golden mussel *(Limnoperrna fortuneiy.)* has advanced rapidly through the Paraná-Tietê waterway network. Populations of this species have caused significant changes to aquatic environments (rivers and reservoirs) and a major economic impact on hydroelectric power stations and sewage systems, due to obstruction and increased corrosion of pipework.

2.4. Waterborne dispersal of exotic species

Since the dawn of mankind, water transport has been vitally important for the exchange of people and goods around the world, moving more than 80 per cent of the planet's goods (ÁGUA DE LASTRO BRASIL, 2009).

Currently, this means of transport accounts for 95 per cent of the volume of products traded in Brazil and, due to the increase in exports, the growth of inland navigation is evidenced by the significant increase in the number of professionals in the merchant navy (SINDMAR, 2010).

Fundamental to the stability of ships, ballast, according to the non-governmental organisation (NGO) Água de Ballro Brasil (2009), is everything that is placed in a ship's hold for this purpose. Solid ballast (stones, clay slabs, wood and sand) was used until the 19th century.

With the modernisation of the nautical industry, the use of water has become widespread, collected by pumps in the port of origin, used as ballast and dumped in the port of destination. This practice transfers billions of tonnes of water and sediment every year. Journeys that used to take months are now travelled in a few weeks, so species that were previously unable to withstand the long, time-consuming journeys survive and populate the new environment, which, not infrequently, due to the lack of predators, multiply wildly, causing serious impacts on the native fauna and flora (MINISTÉRIO DO MEIO AMBIENTE, 2003).

Around 12 billion tonnes of ballast water are transferred around the world every year, carrying around 7000 species including viruses, bacteria, invertebrate larvae and eggs, fish, algae and other creatures (MINISTÉRIO DO MEIO AMBIENTE, 2009).

When a species is transported from its habitat to another location, it is considered exotic or introduced (MACK, 1995). By overcoming the adversities of the new environment, such as climatic factors and pollution, two situations can occur: either the species will have to establish itself or, in a few generations, it will become extinct (CARLTON; GELLER, 1993).

However, when the organism shows exponential population growth and manages to disperse, bioinvasion, once established, is difficult to eradicate, especially when it comes to aquatic species (COWIE, 2004). Figure 1 illustrates the stages of ballasting that ships undergo:

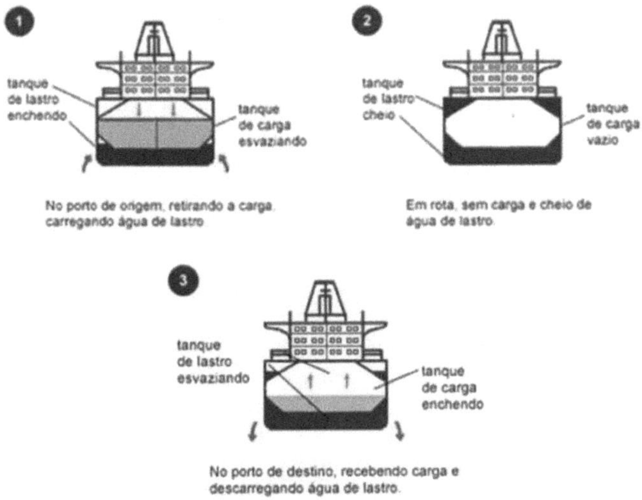

Figure 1 - Ballast water process on ships
Source: LABEC, 2009.

The zebra mussel in North America and the golden mussel in South America are examples of bioinvasion resulting from the disposal of ballast water infested with these organisms at different reproductive stages in ports of destination.

2.5. Legislation to control the spread of exotic aquatic species through maritime transport

The *Globallast* Programme *and* the "International Convention on the Control and Management of Ships' Ballast Water and Sediments" are examples of initiatives by international organisations in the search for solutions to prevent bioinvasives (GLOBALLAST, 2001).

Considered one of the four greatest threats to the oceans and seas, the transfer and introduction of exotic marine organisms into new habitats through ballast water, embedded in the hulls of vessels, has inspired care and the creation of regulations to minimise or prevent infestation in Brazilian ports. Other threats to marine life are pollution, overfishing and habitat destruction (MINISTÉRIO DO MEIO AMBIENTE, 2009).

Since October 2005, the "Maritime Authority Standard for the Management of Ships' Ballast Water" (NORMAN-20/DPC) has been in force in Brazil. In addition to incorporating the provisions of Resolution A.868 of the *International Maritime Organisation* (IMO), it stipulates that all ships, domestic or foreign, with ballast water tanks/ports, using Brazilian ports and terminals must comply with the standards (DIRETÓRIA DE PORTOS E COSTAS, 2005).

Managing ballast water and its fragments is an effective and environmentally safe initiative that not only cuts costs and unnecessary delays for the ship and its cargo, but also reduces risks to the safety of the ship and its crew or the safety of navigation.

The ballast water management plan, according to NORMAN-20/DPC (2005), must be included in the ship's operational documentation and must contain the following items: a) detailed safety procedures for the ship and crew associated with ballast water management;
b) detailed description of the actions to be taken to implement ballast water management;
c) indicate the points where it is possible to take ballast water samples, representative of the ballast the ship is carrying;
d) officer on board responsible for ensuring that the plan is correctly implemented;
e) be written in the working language of the ship; if the language used is not

English, French or Spanish, a translation into one of these languages must be included.

Brazil currently enforces compliance with the most modern standards for managing ballast water on ships, while maintaining its commitment to adapting them as more advanced treatment methods are developed (ÁGUA DE LASTRO BRASIL, 2009).

2.6. Introduction and dispersal of the golden mussel *[Limnopenna fortunei,* Dunker 1857] in South America and Brazil.

The golden mussel, a species from south-east Asia, was brought via the ballast water of ships from Hong Kong or Korea to the Rio de la Plata basin in Argentina, with the first record in 1991 (PASTORINO *et al.,* 1993).

According to studies carried out by Darrigran (2002), this mollusc has been spreading through the South American waterway network at a rate of approximately 240 km/year.

The first record in Brazil was in 1999, in Porto Alegre, Rio Grande do Sul, in the Guaíba River Basin. In 2001, through the Paraguay-Paraná and Paraná-Tietê waterway system, it reached the states of Mato Grosso, Mato Grosso do Sul, Paraná, São Paulo, Minas Gerais through the Rio Grande, Goiás, as well as Rio Grande do Sul, reaching important hydroelectric power stations both in Brazil and in neighbouring countries.

Due to its high reproductive capacity and ability to settle in a wide variety of substrates, it has caused serious environmental and economic damage to fishing activities, hydroelectric power stations, water supply and treatment companies (BARBOSA, 2009).

The dispersal of the golden mussel followed the following route between 1991 and 2004:

a) 1991 - Bagliardi Spa (PASTORINO *et* al. 1993);

b) 1994 - Uruguayan coast of the River Plate, Colonia de Sacramento (SCARABINO; VERDE, 1994);

c) 1995 - Uruguay River (DARRIGRAN; DRAGO, 2000);

d) 1996 - Uruguay River (DI PÉRSIA; BONETTO, 1997);

e) 1997 - Port of Asunción Paraguay River (DARRIGRAN; DRAGO, 2000);

f) 1998 - Port of Posadas on the Paraná River (DARRIGRAN; DRAGO, 2000);

g) 1998 - Upper Paraguay River Basin (CALLIL; MANSUR, 2002);

h) 1999 - Itapuã Beach and Porto das Pombas Municipality of Viamão RS (MANSUR et al., 1999);

i) 1998 - Uruguay River (DRAGO et *al.,* 1998);

j) 2001 - Itaipu Power Station (ZANELLA; MARENDA, 2002);

k) 2002 - Municipality of Rosana (AVELAR et *al.,* 2004);

l) 2003 - São Simão Rio Paranaíba Power Station (BELZ, 2006);

m) 2004 - Barra Bonita Power Station Rio Tietê AES Tietê (BELZ, 2006).

Figure *L* shows the dispersal history of *Limnoperna fortunei* in South America:

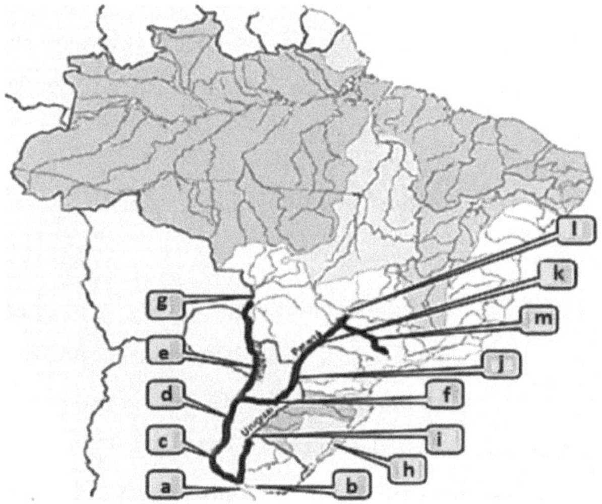

Figure 2 - Location of mussels in South America from 1991 to 2004

Source: BELZ, 2006.

In the Miranda river sub-basin, one of the main tributaries of the river Paraguay, *Limnoperna fortunei* has only been recorded in the lower stretch of the river. In 2003 it was recorded in the Passo do Lontra region (Mato Grosso do Sul) (BARROS et al., 2003).

At hydroelectric power stations (Porto Primavera and Sérgio Motta) downstream on the Paraná River in São Paulo, golden mussels have been reported in the municipality of Primavera (AVELAR et *al.,* 2004).

In the Miranda River sub-basin, one of the main tributaries of the Paraguay River, *Limnoperna fortunei* was only recorded in the lower stretch of the river.

River. In 2003 it was recorded in the Passo do Lontra region (Mato Grosso do Sul) (BARROS *et al.,* 2003).

In September 2004, golden mussels were observed 15 km upstream from the mouth of the Vermelho River (Minas Gerais), which indicates that they are dispersing and may reach the Aquidauana River (Mato Grosso do Sul) (BARROS et al., 2003).

In 2004, the golden mussel was observed in the Apa River, at the southern end of the Upper Paraguay river basin. Its presence has also been investigated in the Cuiabá River, near its mouth with the Paraguay River (OLIVEIRA, et al., 2004).

2.7. Characteristics of the golden mussel *Limnoperna fortunei* (Dunker1857)

A bivalve mollusc from the freshwater Mytilidae family, the golden mussel *(Limnoperna fortunei*, Dunker, 1857) is a prolific animal that forms colonies whose population density varies between 100,000 and 140,000 individuals per m^2 (square metres) (MANSUR et *al.*, 2003; SANTOS et *al.*, 2005) that adhere to various types of surfaces, both natural and artificial, as found in the study carried out by Mansur et *al.,* (2009). Table 1 shows the classification of the species, according to Newell (1969):

Table 1 - Classification of the species *Limnoperna fortunei*

FILO:	Mollusca
CLASS:	Bivalvia
SUBCLASSE:	Pteriomorpha
ORDER:	Mytiloida
SUPERFAMILY:	Mytiloidea
FAMILY:	Mytilidae
GENDER:	Limnoperna
SPECIES:	*Limnoperna* fortunei (Dunker, 1857)

Source: NEWELL, 1969.

The shell of the golden mussel is extremely thin and extended, with a rounded posterior end and a pointed anterior end, made up of two valves (right and left) joined by a ligament on the back inside which the animal's body is protected, as shown in Figure 3:

Figure 3: Golden mussel with the protection of the bivalve shell (bottom) and without the bivalve shell (top)

Source: SILVA, 2010; CHIARIONI, 2011.

Generally, the two valves are morphologically similar and can reach up to 4 cm (RUPPERT; BARNES, 2004). Sexual maturity is reached at approximately 30 days of age with an average size of 0.5 mm (COLARES, 2002).

Life expectancy varies according to geographical location, reaching up to 4.5 years in central China (MAGARA *et al.,* 2001).

The strong attachment to natural and artificial solid substrates is due to the secretion of bisso through a gland located at the base of the muscular foot (MORTON et *al.,* 1998). This material consists mainly of collagen hydrophilic polyphenols that confer adherence (BELZ, 2006).

Through this substance, it adheres and settles, covering large areas and forming colonies that clog up pipes, filters and irrigation channels.

It also gets embedded in boat hulls, harbour structures, pipes and the cooling systems of small boats. Due to its thin, elongated shell, the animal tears fishing nets when they come into contact with colonies attached to rocks and algae (GUIMARÃES, 2006). Figure 4 illustrates the high concentration of individuals in a golden mussel colony:

Figure 4 - High population density of a golden mussel colony
Source: ITAIPU, 2010.

In adulthood, it is resistant to various pollutants and low oxygenation. It is a filter animal with bioaccumulator potential (CATALDO; BOLTOVSKY, 2000).

The ideal temperature for the golden mussel is between 17°C and 30°C and it tolerates salinity of up to 3% (BOLTOVSKY; CATALDO, 1999; MAGARA *et al*., 2001 and CAPITOLI; BEMVENUTI, 2004).

The yellowish colour of the shell, depending on the biotic and abiotic factors of the place where the individual developed, can vary to black and the inside of the shell is pearly (KIMURA, 1994).

The colonies have a greater number of females (around 2/3 of females). Spawning, in regions where the water temperature is higher, can occur several times during the year, and thousands of eggs and larvae settle, increasing the macro fauna or forming new colonies (DARRIGRAN *et* al., 1998).

Studies carried out by researchers Cataldo and Boltovsky (2000) suggest that in the northernmost waters of the Paraná River basin, spawning occurs continuously over nine months of the year, with a significant increase in the warmer periods.

However, it is likely that in locations where the water temperature is relatively constant and higher throughout the year, reproduction takes place continuously.

Because it is a dioecious species, the golden mussel *(Limnoperna fortunei)* has external fertilisation, and the spermatozoa are already productive throughout the annual period (DARRIGRAN et *al.,* 1998).

The duration of the larval stages is directly proportional to the temperature of the water where the colony is established, and the periods and measurements of the stages shown in figure 4 are based on experimentally determined values at 25°C (KIMURA; SEKIGUCHI, 1996).

According to experiments carried out by Kimura and Sekiguchi (1996), the non-valved larval stage ranges from the egg to the fourth-stage trochophore:

Egg: was not observed in this sample, but according to Choi and Shin (1985) apud Santos et *al.*, (2005), it has a cylindrical shape, surrounded by viscous material and is about 80pm long.

Ciliated larva: Presence of small cilia on the outer surface, and the larva measures approximately 80pm in diameter.

Trochophore - Stage 1: The shape is no longer cylindrical, leading to a more elongated shape, acquiring the form of a trochophore type larva. It has a crown of cilia that promote the larva's locomotion. Stage 1 is 80pm to 100pm long and 85pm wide.

Trochophore - Stage 2: The tapered shape becomes wider, leading to a pear-shaped structure, with a slight cavity at the oral end where the mouth is located. At this stage the larva measures 1 lOpm in length

Trochophore - Stage 3: There is a considerable increase in its structure, reaching 125pm in length and approximately 100pm, there is a lateral flattening where the larval shell develops.

Trochophore - Stage 4: There is no increase in size, but the first traces of shell growth appear, where there are changes in its structure due to the shell growing in lateral flattening.

The non-valved larval stage ranges from the "D" shaped larva to the Pedivelliger. The stages of this phase are characterised by their shape, size and the development of the hinges of the larval shells.

D" larva - The "D" represents the shape in which the shell develops in the larva, the hinge (the junction between the two valves) is straight. There is a timid veil in formation, not fully emerging from the edges of the shell. Its length reached between 100 pm and 130 pm.

Straight-edged **Veliger** - The veil is already pronounced, its length is a little more than half the length of the valve, growing in a ventral direction. Sharp rounding of the valves, their length varies from 140 pm to 180 pm. The hinge is straight with a slight internal thickening. At this stage the shell is transparent, making it possible to see the larva's internal organs. The larvae move through the water easily and are easily distinguished from plankton. It is characterised by circular movements on its axis. At this stage there is still no foot.

Umbonate veliger - The shell has distinct centred umbos, i.e. slightly elevated. The length varies from 190 pm to 230 pm. The hinge is denser and thicker, the veil well defined and rudiments of the foot can be seen in the ventral part behind the veil. Movement and activity similar to the previous stage.

Pedivellate - With a length of between 240 pm and 270 pm, it reaches its maximum in the larval stage, there is mobility by crawling, due to the functional foot, it maintains its swimming characteristics, in a circular way, with the aid of the veil.

Post-larva stage - Movement with the foot helps them to attach to the substrate, with the help of the bisso. The set of filaments called bisso often detaches, reattaching to substrates close to the original location.

The juvenile stage of the golden mussel is around 30 days old and 5 millimetres (mm) long, where they are able to reproduce (COLARES *et aL,* 2002). Figure 5 shows the phases mentioned in relation to the non-valved larval stage.

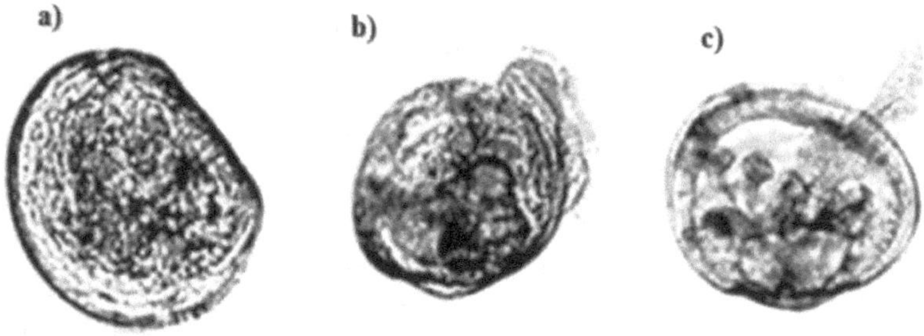

Figure 5 - *Limnoperna fortunei* larval stages. a) D larva; b) umbonate larva; c) pedivellate larva.
Source: CATALDO and BOLTOVSKOY, 2000.

The high population density of mussels also rapidly modifies the presence and abundance of various species of native organisms. With less prey available, some fish feed on the mussels, but many of them are unable to digest the shells and die. Figure 6 shows the fish species *Pterodoras granulosus* which revealed, in addition to spawn and detritus, a high quantity of golden mussels *(Limnopema fortunei),* ingested by the specimens (Figure 6):

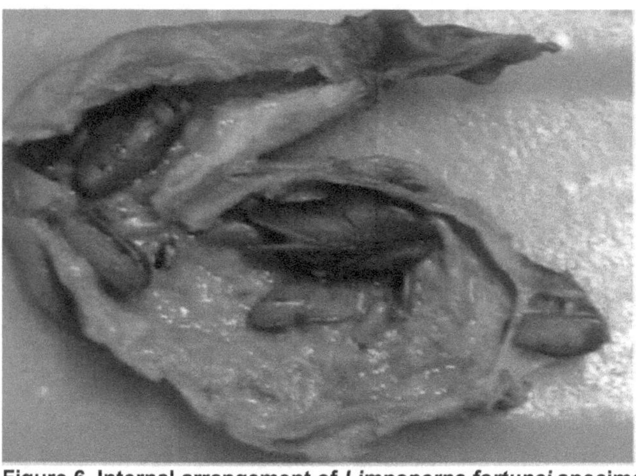

Figure 6. Internal arrangement of *Limnoperna fortunei* specimens in the stomach of the armai, *Pterodoras granulosus.*
Source: VERMULM JÚNIOR et al., 2008.

In addition, the golden mussel has the highest filtration rate (which represents the demand for food) among invasive bivalves, which also includes the zebra mussel, one of the most impactful in North America and Europe. Considering the population contingents, the invasion by the golden mussel demands a large volume of plankton for its food, which impacts the entire food chain in the region and can reduce the populations of fish and other organisms in the ecosystem that depend mainly on plankton as a food source (SILVA, 2016).

Von Rúckert and collaborators (2004) demonstrated through research carried out in the Paraná River that the golden mussel feeds on the cyanobacterium *Microcystis viridis*, which poses a great risk of cyanotoxin bioaccumulation, since it is a highly toxic species.

According to Darrigran (2002), it is a species that is resistant to adverse conditions, as shown in Table 1, a fact that, combined with the absence of predators, can be attributed to the bioinvasor's success on the South American continent.

Table 1 - Physico-chemical survival limits for *Limnoperna fortunei.*

Parameters	Limits
Salinity	0-12%
Calcium	3.96 mg-L^{-1}
pH	6,2-7,4
Temperature of larval development	16°C-28°C
Adult survival temperature	8°C - 35°C
Oxygen	< 1.0 mg-L^{-1}
Exposure to air	Up to 7 days

Source: DARRIGRAN, 2002.

2.8. Golden mussel control measures

The advance of the golden mussel has led to intense monitoring to ensure that other rivers are not affected by this bioinvasor. One of the main channels for the dispersal of *Limnoperna fortunei is* through commercial shipping between Argentina and other South American countries, where larvae and adults are attached to the hulls and water intake systems of vessels (OLIVEIRA et *al.,* 2000).

Figure 7 illustrates golden mussel encrustation on some types of substrate such as stone, bricks, wood and metal.

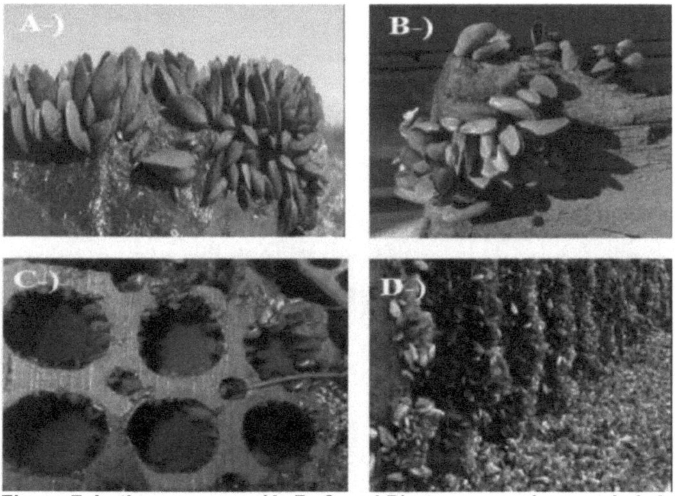

Figure 7. In the sequence (A, B, C and D) stone, wood, ceramic brick and metal with golden mussel attached.

Source: FURNAS, 2005.

Some recommendations to prevent new proliferations of the golden mussel in the rivers belonging to the Upper Paraguay river basin are cited in studies carried out by Oliveira *et al.* (2004):

1. Monitor and provide updated information on the area of occurrence of *Limnoperna fortunei* in the Upper Paraguay river basin;

2. Check boats and motors transported by land for the presence of fouling on the outside, and remove water debris and plants from areas where golden mussels occur;

3. Do not return mussels removed during the cleaning process to the aquatic environment or drainage channels; dispose of them on land;

4. Do not transfer fishing material from the Paraguay River, or other places where golden mussels occur, to fish farming ponds, so as not to contaminate the crops;

5. Do not transfer any type of material originating in the Paraguay River, or elsewhere where the golden mussel is present, to rivers and streams where

the mussel does not occur;

6. Do not dispose of water from containers containing live bait in water bodies or drains, dispose of them on land.

Barbosa (2009) used the ground residue of the golden mussel as a soil acidity corrector and a source of nutrients for plants, with satisfactory results in correcting the soil and supplying part of the nitrogen and phosphorus requirements for growing maize.

[2]Silva (2010) and Chiarioni (2011) carried out work to use ground golden mussel waste in agriculture, as a fertiliser in association with vinasse and as a soil acidity corrector in sugar cane plantations.

2.9. Environmental and economic issues

The flow of boats and ferries through the waterway network has been considered the biggest vehicle for dispersing the mollusc throughout Argentina and other South American countries. Studies on the presence of the animal in Brazilian rivers suggest that larvae and individuals already in the reproductive stage attach themselves to the hulls and water catchment systems of boats and thus travel long distances (OLIVEIRA *et al.,* 2000).

Figure 8 illustrates fouling on the grilles of water intakes at hydroelectric power stations, causing a loss of system load and reducing performance, as well as causing economic damage to hydroelectric power stations.

[2]The work was carried out by the authors of the book during their undergraduate studies in Biofuels Technology at the Araçatuba Faculty of Technology (Fatec) between 2010 and 2011.

Figure 8 - Hydroelectric plant grid colonised by golden mussels
Source: FURNAS, 2005.

Studies carried out by Darrigran and colleagues (1998) on the composition of the fauna of hard substrates colonised by golden mussels show a significant increase in the richness and abundance of groups of non-mollusc invertebrates such as *Oligochaeta, Hirudinea, Isopoda and Chironomidae.*

However, there has been a significant impact on native mollusc species gastropods, as evidenced by the large reduction in diversity and number of species, which has been exacerbated by competition for food and the absence of predators to control the population of the invasive bivalve. Figure 9 shows mussels colonising the shell of *Anodontites trapezialis, a* bivalve species native to South America.

Figure 9 - Colony of golden mussel on the shell of the native bivalve mollusc *Anodontites trapezialis*

Source: BELZ, 2006.

The golden mussel has been found attached to algae, plant roots and native aquatic animals, damaging the structure to which it adheres, competing for space and food (DARRIGRAN, 2000; MANSUR *et al.*, 1999; MANSUR et *al.*, 2003). Like most filter-feeding bivalve molluscs, it accumulates pollutants present in the environment (VILLAR, 2006). The presence of a species that bioaccumulates polluting compounds and is capable of forming large colonies can promote an increase in biomagnification processes.

Biomagnification is the process by which a substance is absorbed by an aquatic organism through the food chain and presents increasing concentrations in various organisms, according to their trophic level. This can cause problems for the consumption of aquatic organisms, including fish, and seriously affect economic activities associated with fishing (BELZ, 2006).

Economically, this mollusc causes a great deal of damage. According to Phillips et al. (2005), the extensive proliferation and settlement of the golden mussel in hydroelectric power stations can mainly cause:

a) clogging or reduction of the pipe section;
b) decomposition of organic material;
c) increased corrosion of pipes, metal alloys, concrete and polymers, due to

the proliferation of other undesirable biological agents forming biofilms (bacteria, fungi, etc.);

d) Decreased service life of equipment due to increased handling during maintenance;

e) increased labour costs for cleaning or replacing pipes, filters, etc;

f) reduction in the speed of water flow in pipes due to friction losses (turbulent flows);

g) accumulation of empty valves and contamination of waterways by mass mortality due to inadequate control treatments;

h) filter occlusion;

j) reduced efficiency of heat exchange equipment.

The macro fauna formed in hydroelectric plant pipelines can be seen in Figure 10:

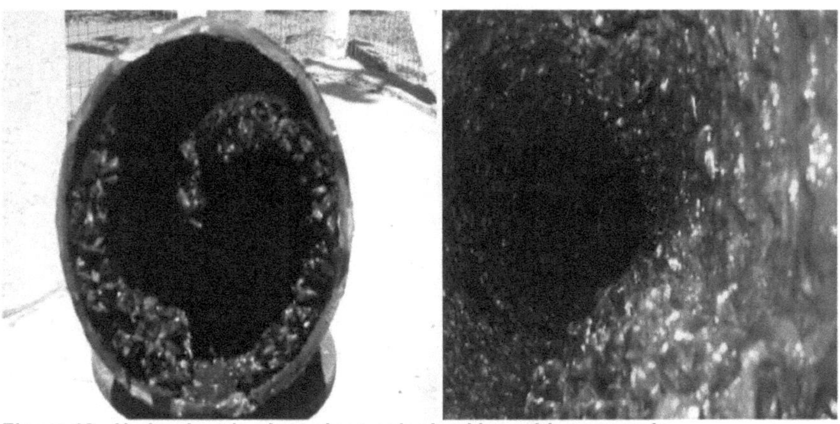

Figure 10 - Hydroelectric plant pipes colonised by golden mussels
Source: PHILLIPS *et al.*, 2005

Water pump motors used in hydroelectric plant reservoirs need constant maintenance due to the formation of golden mussel colonies, as shown in Figure 11.

Figure 11 - Golden mussels encrusted in a water pump motor
Source: ITAIPU, 2002.

CHAPTER 3

MATERIALS AND METHODS

The methodology used was *on-site* data collection and a bibliographical survey on the proliferation of the golden mussel in the city and region of Araçatuba between 2010 and 2015.

The golden mussel *(Limnoperna fortunei)* was first observed in August 2010 on a floating platform located at the *Yacht Club* Araçatuba, on the waters of the River Tietê.

Using a BAK *Japan* BK-GPS4306® global positioning system (GPS) device, with the following location: South Latitude 21.04792° and West Latitude 50.51692°, we observed the golden mussel colonies attached to the floating platform. Figure 12 shows the satellite image obtained using *Google Earth software,* indicating the location where the golden mussel was observed (circled area).

Figure 12 - Satellite image of the Yacht Club Araçatuba's floating platform on the left bank of the River Tietê

SOURCE: GOOGLE EARTH, 2011.

According to (Mansur, 2003) golden mussels form large colonies known as *macrofouling* (macrofaunas), reaching populations of 100,000 to 140,000-

m^{-2} individuals.

Its bodily constitution is characterised by two shells connected to each other by the back (bivalve), protecting its body and its colour varies between bright green *and* yellowish brown (PASTORINO *et al.,* 1993).

Figure 13 shows the structure of the floating platform with golden mussel *macrofouling*.

Figure 13. Floating platform and the proliferation of golden mussels on its structure
Source: PERSONAL ARCHIVE.

Figure 14 shows the structure of the Rio Tietê Shipyard, in Araçatuba/SP, used for the waterway transport of ethanol, where its submerged structure, after being stopped for maintenance, showed golden mussel *macrofouling* along its entire length.

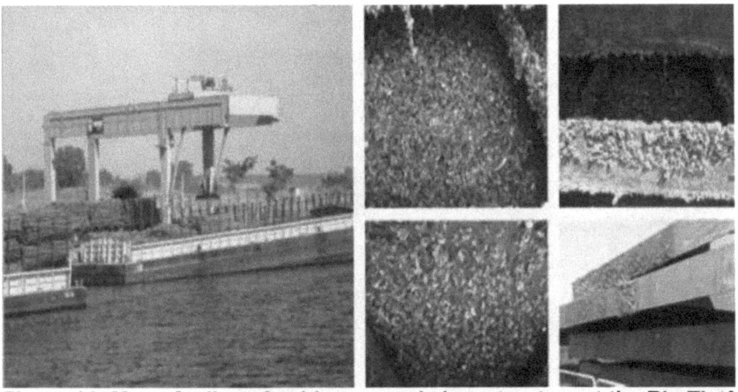

Figure 14. *Macrofouling* of golden mussels in a structure at the Rio Tietê Shipyard in the Araçatuba/SP region.

Source: CHIARIONI *et al.*, 2011, ATOGUIA *et al.*, 2013.

Confirming studies carried out by Brandy (2001), the golden mussel has a great ability to attach itself to structures made of wood and metal in the waterway harbours of Araçatuba/SP, forming *macrofouling* of varying population densities all over the surface (MANSUR *et al.*, 2003).

Because it is submerged in the Tietê River for a large part of the time, the golden mussel can cause environmental impacts on the food chain of native filter feeders, as well as contributing to the increased proliferation of aquatic macrophytes caused by the decrease in the river's turbidity (VON RÚCKERT, et *al.*, 2004).

The Tietê River has ideal characteristics for the development of the golden mussel, such as temperatures between 17°C and 30°C and high concentrations of phytoplankton biomass, with favourable conditions for the development of the invasive bivalve (SILVA, 2006).

As well as damaging the performance of boats due to golden mussel fouling in engines and pipes, reducing the useful life of equipment, it also causes serious damage to hydroelectric power stations (RESENDE, 2007).

In the regions close to the River Tietê it is possible to find golden mussels *macroufouling* immersed in substrates, even using aquatic macrophytes in competition for space in the environment. Figure 15 shows the interaction between the golden mussel and aquatic macrophytes.

Figure 15. ***Macrofouling*** **of mussels using aquatic macrophytes as a substrate in the Tietê River in the region of Araçatuba/SP.**

Source: FOLHA DA REGIÃO (2012).

Figure 16 shows a car that was submerged in the waters of the River Tietê after a long dry spell in 2014, with colonies of golden mussels on its internal and external structure.

Figure 16. ***Macrofouling*** **of golden mussels in a vehicle that was immersed in the Tietê River in the Araçatuba/SP region.**

Source: G1 (2014).

The golden mussel jeopardises fish farms in the Araçatuba/SP region. The mollusc uses the nets in the tanks as a substrate, damaging the development and growth of the fish. Figure 17 shows pond nets with golden mussel *macrofouling*.

Figure 17. Golden mussel colonies in nets in fish breeding ponds on the Tietê River in the Araçatuba/SP region.

Source: G1 (2014).

A working group formed to solve the invasion of the golden mussel in the Ilha Solteira reservoir, in the Araçatuba region, has classified the eradication of the species as very difficult. The experts have decided that efforts will focus on preventing the mollusc from spreading to other rivers in the country.

Figure 18 shows the golden mussel being questioned by the Public Prosecutor's Office, which in 2012 filed public civil actions relating to the Ilha Solteira and Água Vermelha hydroelectric power stations, requesting measures to eradicate the golden mussel.

Figure 18. The proliferation of the golden mussel being questioned by the Public Prosecutor's Office at hydroelectric power stations.
Source: FOLHA DA REGIÃO (2015).

With the information collected, an information poster was created to be used in the classroom about bioinvasors in nature, as proposed in the learning situation for the students. As electronic equipment is not allowed in the Penitentiary Progression Centre, the poster was used as a strategy to promote environmental education to the students.

Based on the study of bioinvasors related to the complementary learning situation proposed in the CEEJA (State Centres for Youth and Adult Education) material, Unit 1 - Interdependence of Living Beings makes students reflect on the contextualisation of environmental preservation (CEEJA, 2016).

Students are able to argue procedurally and attitudinally about the problems that can be caused by bioinvasors. Citizenship is encouraged in learning. Knowing that we have a responsibility to preserve nature, to ensure that natural resources are able to maintain the necessary balance for future generations.

By establishing a fact that is happening on the subject of bioinvasion to the students, learning that is contextualised in everyday life, the results tend to be satisfactory, by providing a broad debate of ideas in the search for understanding about the environmental interactions around us.

Affectivity considers the intrinsic factors in cognitive development in relation to issues involving the teaching of young people and adults, in order to awaken students' criticality (HERBELE, 2011).

The classrooms at the Penitentiary Progression Centre (CPP) are multigrade, where the students have different levels of knowledge and experience. Depending on the learning situation presented in the subject of natural sciences, students have difficulties in certain aspects where it is necessary to understand the dynamics of species in the environment.

Students present their experiences in relation to what is being developed in the classroom. It favours mutual learning with the teacher, since the assessment contexts will be facilitated to make the concepts developed by the students more effective.

With the data gathered on the golden mussel in a broad context, and on the situation of the rivers located in the region of

Araçatuba/SP, in relation to the presence of the bioinvasor. With the data in hand, a poster was made to be used in the classroom as a teaching tool.

When the topic was addressed, the students were able to get involved in the problem situation, and with the poster available for reading, the debate on what can be done to eradicate the problem was discussed.

As well as realising the depth of the subject, the students had contact with a different structure in relation to investigative scientific reading, how science is changeable and how criticality can establish ideas to create solutions in a social context.

Figure 19 shows the structure of the poster used in the bioinvasors lesson, in the complementary learning proposed in the CEEJA (State Centres for Youth and Adult Education) Unit 1 - Interdependence of Living Beings.

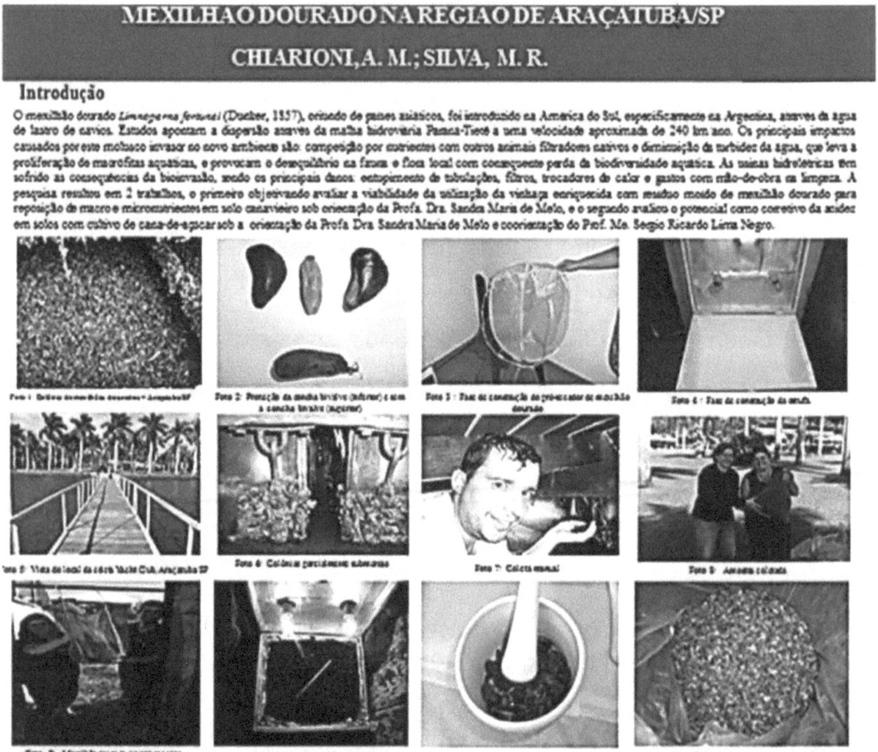

Figure 19. Structure of the poster used for the lecture on the golden mussel

Source: PERSONAL ARCHIVE

The structure of the poster presents images to educate students to rethink their actions and spread the teachings to their community.

CHAPTER 4

RESULTS AND DISCUSSION

Figure 20 shows the initial moment of the lecture to the students using the banner to introduce the subject of bioinvasives in nature, in particular the golden mussel.

Figure 20. Lecture on the bioinvasion of the golden mussel in the Araçatuba/SP region. Source: PENITENTIAL PROGRESSION CENTRE, VALPARAÍSO/SP.

After the presentation of the lecture on bioinvasion in a generalised way, and contextualising learning by showing the students the problem of the golden mussel, there were questions on the subject, promoting debates of ideas. Some of the questions posed by the students are structured in Table 2 (two) below.

Table 2. Questions asked by students after the lecture on the bioinvasion of the golden mussel in the region of Araçatuba/SP

"If you brought the predator of the golden mussel from its country of origin, wouldn't that solve this proliferation?"
"Isn't the golden mussel edible?"
"How did he get to the Tietê River?"
"If you put a chemical in the water, can you fight it?"
"Does he carry diseases with him?"

Source: STRUCTURED BY THE AUTHOR.

Through the students' questions, it was possible to discuss the importance of environmental awareness for the preservation of the planet's ecosystem, as proposed by the learning situation. The pedagogical tool contextualised through the banner brought greater participation and attention to the subject of bioinvasors and their consequences for the environment.

By showing and debating the issue of bioinvasive species and demonstrating the arrival of the golden mussel in the region where the students are deprived of their liberty, there was greater interactivity and reflection on environmental issues.

Environmental education raises environmental awareness and, as a result, learning the content has become more important for the students' role in raising awareness in their community.

CHAPTER 5

FINAL CONSIDERATIONS

The purpose of environmental education is to share and raise awareness through the teaching process of environmental issues and how to bring new alternatives to minimise impacts on the environment.

The construction of thoughts to raise awareness of environmental education is one of the important ways of building a new being in relation to the structures of the environment. With these parameters, the implementation of concepts should be treated as something intrinsic to the human being from the earliest years of life. Throughout their school life, individuals should have compulsory subjects relating to environmental education and sustainable development.

Teaching students about environmental issues is in synergy with other subjects, according to the National Curriculum Parameters (PCN) (BRASIL, 1997). As agents for transforming society and, consequently, the environment in which they live, students will be the basis for raising awareness so that other people understand the need to conserve the planet.

The lecture on environmental education at the Valparaíso Penitentiary Progression Centre (CPP) in the Araçatuba/SP region, in relation to the bioinvasion of the golden mussel, established the work in line with the other teachers from the perspective of meaningful and comprehensive learning in the quest to raise awareness of environmental problems and debate actions to minimise the destructive impacts on nature.

CHAPTER 6

BIBLIOGRAPHICAL REFERENCES

ÁGUA DE LASTRO BRASIL. **Ballast water and its environmental risks.** Available at: <http://www.aguadelastrobrasil.org.br/arquivos/Atividades%20e%20Projetos/ cartilha%20vers%C3%A3o%201.0_portugu%C3%AAs.pdf>. Accessed on: 20 July 2016.

AMBIENTE BRASIL. **Environmental Education.** 2008. Available at:< www.ambientebrasil.com>. Accessed on: 08 July 2016.

ARAÚJO, T. C. A. **Main global milestones in environmental education.** 2007. Available at: <http://noticias.ambientebrasil.com.br/artigos/2007/09/11/33350-principais-marcos-historicos-mundiais-da-educacao-ambiental.html>. Accessed on: 09 July 2016.

ATOGUIA, F. S.; BARIZÃO, G. B.; SANTOS, H. L. **Asian mollusc invades the Tietê River.** In: CICFAL Junior Scientific Initiation Congress. Faculdades Adamantinenses Integradas, 2013. Available at: < http://www.unifai.com.br/cic2017/docs/anais/anais_medio_2013.pdf >. Accessed on: 09 July 2016.

AVELAR, W. E. P.; MARTIM, S. L.; VIANNA, M. P. A. New occurrence of *Limnoperna fortunei* (Dunker 1856) (Bivalvia, Mytilidae) in the State of São Paulo, Brazil. **Brazilian Journal of Biology.** v. 64 n. 4 p. 739-742. 2004.

BARBOSA, D. B. P. **Use of ground golden mussel waste *(Lymnopenna fortunei* Dunker, 1857) as a soil acidity corrector and source of nutrients for plants.** 145 f. Dissertation (Master's in Soil Science), Federal University of Rio Grande do Sul (UFRGS), Porto Alegre, 2009.

BARROS, L. F.; OLIVEIRA, M. D.; SILVA, L. C. R.; EILERS, V., TAKEDA, A.M. Occurrence of the golden mussel *{Limnoperna fortunei)* in the Miranda River, Pantanal-MS, Brazil. **Brazilian Congress of Limnology.** Juiz de Fora, Minas Gerais, 2003.

BELZ, C. E. **Bioinvasion risk analysis by *Limnoperna fortunei* (Dunker, 1857): a model for the Iguaçu River basin, Paraná.** 2006. 102f. Thesis (Doctorate in Zoology), Federal University of Paraná (UFPR), Curitiba, 2006.

BOLTOVSKOY, D.; CATALDO, D. Population dynamics of *Limnoperna fortunei,* an invasive fouling mollusc, in the lower river Paraná (Argentina). **Biofouling,** v. 14. n. 3. p. 255-263. 1999.

BRANDY Jr, R. F. A fracture mechanical analysis of fouling release from nontoxic antifouling coatings. **Progress in Organic Coatings,** v. 43, p. 188192, 2001.

BRAZIL. National Curriculum Parameters. **Introduction to the national curriculum parameters.** Brasília: MEC/SEF, 1997. Available at:< http://portal.mec.gov.br/seb/arquivos/pdf/livro01.pdf>. Accessed on: 10 July 2016.

CADERNO SECAD 1. **Educação Ambiental: aprendizes de sustentabilidade.** 2007. Available at: < http://portal.mec.gov.br/dmdocuments/publicacao2.pdf >. Accessed on: 06 July 2016.

CALLIL, C. T.; MANSUR, M. C. D. Corbiculidae (Mollusca, Bivalvia) in Pantanal of Mato Grosso, upper Paraguay Basin, Brazil: distribution and population density. **Amazoniana,** p. 1-15. 2002.

CAPITOLI, R. R.; BEMVENUTI, C. E. Distribution of the golden mussel *Limnoperna fortunei* (Dunker, 1857) in the estuarine area of the Patos Lagoon and São Gonçalo Channel. **Proceedings: IV Symposium on Brazilian Ecosystems**, São José dos Campos, 2004.

CARLTON, J. T.; GELLER, J. B. Ecological roulette: biological invasions and the global transport of nonindigenous marine organisms. **Science,** n. 26, p. 78-82, 1993.

CATALDO, D. H.; BOLTOVSKY. D. Yearly reproductive activity of *Limnoperna fortunei* (Bivalvia) as inferred from the occurrence of its larvae in the plankton of the lower Paraná river and the Rio de la Plata estuary (Argentina). **Aquatic Ecology**, Dordrecht, Netherland, v. 34 p. 307-317, 2000.

CEEJA (State Centres for Youth and Adult Education). **Biology Volume 1 - Interdependence of Living Beings (Unit 1).** 2016. Available at:< http://www.ejamundodotrabalho.sp.gov.br/ConteudoCEEJA.aspx?Material D= 71&tipo=Aluno>. Accessed on: 02 July 2016

CHIARIONI, A. M. **Study of the feasibility of using ground golden mussel waste *(Limnopenna fortunei,* Dunker, 1857) as an acidity corrector in sugarcane soil.** Araçatuba, São Paulo, 2011. Monograph (Degree in Biofuels Technology) - Faculty of Technology

(Fatec).

CHIARIONI, A. M.; SILVA, M. R.; MELO, S. M. **Occurrence and impacts caused by invasive bivalve in the region of Araçatuba-SP.** In: CICFAI. Congress of Scientific Initiation. Faculdades Adamantinenses Integradas, 2011. Available at: < http://www.fai.com.br/cicfai2015/docs/anais/anais_biologicas_2011. pdf >. Accessed on: 07 July 2016.

COLARES, E. R. C. Diagnosis and control of the golden mussel, *Limnoperna fortunei*, in water treatment systems in Porto Alegre. In: **Italian-Brazilian Symposium on Sanitary and Environmental Engineering.** Vitória, Espírito Santo, 2002.

COSTA, L. A. V.; IGNÁCIO, R. P. **Consumer relations x the environment: in search of sustainable development.** 2014. Available at: <http://www.ambito-juridico.com. br/site/?n_link=revista_artigos%20_leitura&artigo_id=10794&revista_caderno=5>. Accessed on: 08 July 2016.

COWIE, R. H. Disappearing snails and alien invasions: the biodiversity/conservation interface in the pacific. **Journal of Conchology.** Special Publication 3, p. 23-37, 2004.

DARRIGRAN, G. A.; DAMBORENEA, M. C.; PENCHASZADEH, P. E. A case of hermaphroditism in the freshwater invading bivalve *Limnoperna fortune* (Dunker, 1857) (Mytilidae) from Rio de la Plata, Argentina. **Iberus.** v. 16, p. 99-104. 1998.

DARRIGRAN, G.; DRAGO, I. E. Ivasion of the exotic freshwater mussel *Limnoperna* fortunei (Dunker, 1857) (Bivalvia: Mytilidae) in South America. **The Nautilus,** p. 69-73. 2000.

DARRIGRAN, G. A. Invasive freshwater bivalves of the neotropical region. **Dreissena,** v. 2, n. 2, p. 7-13, 2002.

DI PÉRSIA, D.; BONETTO, A. A. 1997. New *Limnoperna fortunei* cites for the Paraná River basin, Argentina. **Neotropica.** p. 119-120. 1997.

DIRECTORATE OF PORTS AND COASTS (DPC). **NORMAM 20/DPC.** Available at: <https://www.dpc.mar.mil.br/normam/N_20/N_20.htm>. Accessed on 10 July 2016.

DRAGO, I. E.; DARRIGRAN, G.; SCARABINO, F.; OLIVEROS, O.B. Update on the distribution of *Limnoperna fortunei* (Bivalvia, Mytilidae): Uruguay River and some tributaries. In: **V Latin American Congress of Ecology,** 1998.

REGIONAL BROADSHEET. **Research seeks alternative to golden mussel bioinvasion.** 2012. Available at: <http://www.folhadaregiao.com.br/Materia.php?id=297489>. Accessed on: 07 July 2016.

REGIONAL BROADSHEET. **Group from Ilha Solteira tries to prevent dispersal of the golden mussel.** 2015. Available at: <http://www.folhadaregiao.com.br/Materia.php?id=379561>. Accessed on: 08 July 2016.

FURNAS. **Booklet on the Golden Mussel.** Available at:< http://www.furnas.com.br/arcs/pdf/Mexilh%C3%A3oDouradoCartilha.pdf>. Accessed on: 13 July 2016.

G1. **Car covered in algae emerges as the Tietê River dries up.** 2014. Available at: <http://g 1 .globo.com/sao-paulo/sao-jose-do-rio-preto-aracatuba/noticia/2014/08/com-seca-no-tiete-car-submerso-e-encontrado-coberto-por-algas.html>. Accessed on: 07 July 2016.

G1. **Disorganised mussel proliferation creates a problem in the river Grande.** 2014. Available at:< http://g1.globo.com/sao-paulo/sao-jose-do-rio-preto-aracatuba/noticia/2014/07/proliferacao-desordenada-de-mexilhao- gera-problemas-no-rio-tiete.html>. Accessed on: 07 July 2016.

GALLI, A. **Environmental education as a tool for sustainable development.** Maringá, 2007. Dissertation (Master's Degree in Economic and Social Law) - Pontifical Catholic University of Paraná.

GLOBALLAST. **1st International Ballast Water Treatment Standards Workshop.** IMO London. March 2001 Workshop Report. Available at:< http://globallast.imo.org/workshopreport.htm>. Accessed on: 20 July 2009.

GUIMARÃES, C. **Water carried as ballast in the hold of ships travelling the oceans carries organisms capable of generating epidemics and ecological imbalance.** 2006. Available at:< http://www.radiobras.gov.br/especiais/meioambiente_navios/poluicaonavios _caoa.htm>. Accessed on: 14 July 2016.

HERBELE, K. **Importance and use of playful activities in the education of young people and adults.** 44p. Course Conclusion Work for Specialist in Professional Education Integrated with Basic Education in the EJA Modality, Federal Technological University of Paraná, 2011.

INEP - NATIONAL INSTITUTE FOR EDUCATIONAL STUDIES AND RESEARCH ANÍSIO TEIXEIRA. **Inclusion of Science in Saeb: basic document.** Brasília: Anísio Teixeira National Institute for Educational Studies and Research, 2013.

ITAIPU. **Meeting discusses golden mussel control.** 2002. Available at: <http://jie.itaipu.gov.br/print_node.php?secao=turbinadas1&nid=2532>. Accessed on: 11 July 2016.

JANKE, N. **National environmental education policy:** contradictions and disputes. Bauru, 2012. Thesis (Doctorate in Science Education) - Faculty of Sciences, São Paulo State University - UNESP.

KIMURA, T. The earliest record of *Limnoperna fortunei* (Dunker, 1857) from Japan. **Chirobatan,** p. 34-35, 1994.

KIMURA, T.; SEKIGUCHI, H. Effects of temperature on larval development of two mytilid species and their implication. **Venus. Jap. J. Malacol.** p. 215222. 1996.

LABEC. **Ballast water.** Available at: < http://www.labec.com.br/biodigital/agua-de-lastro/>. Accessed on: 20 July 2016.

MACHADO, C. J. S.; OLIVEIRA, A. E. S. **Invasive alien species:** a **national problem that is still little known.** 2009. Available at:< http://cienciaecultura.bvs.br/pdf/cic/v61n1/a10v61n1.pdf>. Accessed on: 11 July 2016.

MACK, R. N. Understanding the processes of weed invasions: the influence of environmental stochasticity. In C. Stirton, editor. Weeds in a changing world. **British Crop Protection Council, Symposium Proceedings,** n. 64. p. 65-74. Brighton, UK, 1995.

MAGARA, Y.; MATSUI, Y.; GOTO, Y.; YUASA, A. Invasion of the non-indigenous nuisance mussel, *Limnoperna fortunei*, into water supply facilities in Japan. **Aqua Journal of Water Supply: Research and Technology,** p. 113-124, 2001.

MANSUR, M. C. D.; RICHINITTI, L. M. Z.; SANTOS, C. P. *Limnoperna fortunei* (Dunker, 1857), an invasive bivalve mollusc, in the Guaíba basin,

Rio Grande do Sul, Brazil. **Biociências,** p. 147-150, 1999.

MANSUR, M. C. D.; SANTOS, C. P.; DARRIGAN, G.; HEYDRICH, I.; CALLIL. C. T.; CARDOSO, F. R. First quali-quantitative data of the golden mussel, *Limnoperna fortunei* (Dunker), in the Jacuí Delta, Guaíba Lake and Patos Lagoon, Rio Grande do Sul, Brazil and some aspects of its invasion in the new environment. **Revista Brasileira de Zoologia,** v. 20, n. 1, p. 75-84, 2003.

MMA - Ministry of the Environment. **Communication project - golden mussel.** Support document for communication actions. Brasilia. 2003.

MMA - Ministry of the Environment. **The different shades of environmental education in Brazil.** 2009. Available at: <http://www.mma.gov.br/estruturas/educamb/_arquivos/dif_matizes.pdf>. Accessed on: 10 July 2016.

MMA - Ministry of the Environment. **Invasive alien species.** 2010. Available at: < http://www.mma.gov.br/biodiversidade/biosseguranca/especies-exoticas-invasive>. Accessed on: 08 July 2016.

MMA - Ministry of the Environment. **National strategy on invasive alien species.** 2011. Available at: <http://www.mma.gov.br/estruturas/174/_arquivos/anexo_resoluoconabio05_ national_strategy invasive ________________ invasive_species_annex_resoluoconabio05_174.pdf>. Accessed on: 11 July 2016.

MMA - Ministry of the Environment. **National report on invasive alien species.** Available at: <http://www.mma.gov.br/biodiversidade/biosseguranca/especies-exoticas-invasoras/informe-nacional>. Accessed on: 11 July 2016.

MORTON, B.; PREZANT, R. S.; WILSON, B. Class Bivalvia. In: BEELEY, P. L.; ROSS, G .J. B.; WELLS, A. **The Souther Synthesis. Fauna of Australia.** Vol. 5, ed. Melbourne, 1998.

NEWELL, N. D. Classification of Bivalvia. In: R. Moore (Ed.). **Treatise on Invertebrate Palaeontology.** Lawrence, University of Kansas, Part N, vol. 1, p. 205-223, 1969.

OLIVEIRA; M. D.; TAKEDA, A. M.; BARBOSA, D. S.; CALHEIROS, D. F. Occurrence of the exotic species *Limnoperna fortunei* (Bivalvia, Mytilidae) in the Paraguay River, Pantanal, Brazil. In: **Symposium on the Natural and**

Socio-Economic Resources of the Pantanal. Corumbá: Embrapa Pantanal, 2000.

OLIVEIRA, M. D.; PELLEGRIN, L. A.; BARRETO, R. R.; SANTOS, C. L.; XAVIER, I. G. Area of occurrence of the golden mussel *(Limnoperna fortunei)* in the Upper Paraguay river basin, between 1998 and 2004. **EMBRAPA Documents, Corumbá: EMBRAPA Pantanal**, 2004.

PASTORINO, G.; DARRIGRAN, G.; MARTÍN, S. M.; LUNASCHI, L. *Limnoperna fortunei* (Dunker, 1857) (Mytilidae), a new invasive bivalve in the waters of the Rio de La Plata. **Neotrópica,** v. 39 p. 34, 1993.

PHILLIPS, S.; DARLAND, T.; SYTSMA, M. **Potential economic impacts of zebra mussels on the hydropower facilities in the Columbia river basin. prepared for the Bonneville power administration by Pacific States Marine Fisheries Commission.** Available at:< http://www.psmfc.org/>. Accessed on: 27 Nov. 2010.

PRS - Solid Waste Portal. **A brief history of global environmental education.** 2013. Available at:< http://www.portalresiduossolidos.com/historia- da-educacao-ambiental-brasil-e-mundo/>. Accessed on: 10 July 2016.

RESENDE, M. F. **The variation of hydraulic characteristics in forced conduits due to infestation by *Limnoperna fortunei.*** 101p. Dissertation (Master's Degree) Federal University of Minas Gerais, 2007.

RUPPERT, E. E.; BARNES, R. D. **Zoologia dos invertebrados.** Ed. 6, p. 353-484, São Paulo, 1996.

SANTOS, C. P.; WÚRDIG, N. L.; MANSUR, M. C. D. Larval stages of the golden mussel *Limnoperna fortunei* (Dunker) (Mollusca, Bivalvia, Mytillidae) in the Guaíba basin, Rio Grande do Sul, Brazil. **Brazilian Journal of Zoology.** Ed. 3, v. 3, p. 702-708, 2005.

SCARABINO, F.; VERDE, M. *Limnoperna* fortunei (Dunker, 1857) in the Uruguayan coast of Rio de la Plata (Bivalvia; Mytilidae). **Malacological Society of Uruguay,** p. 374-376. 1994.

SENATE IN DISCUSSION. **From the United Nations Conference on the Human Environment in Stockholm to Ri o-92:** environmental agenda for countries and drafting of documents by the World Commission on Environment and Development. 2014. Available at:< http://www.senado.gov.br/noticias/Jornal/emdiscussao/rio20/a-rio20/co.

aspx>. Accessed on: 10 July 2016.

SILVA, D. P. **Bioecological aspects of the golden mussel *Limnoperna fortunei* (Bivalvia, Mytilidae) (Dunker, 1857).** 138p. Thesis (Doctorate) Federal University of Paraná, 2006.

SILVA, F A.; NEWTON P. U.; BARBOSA, R. S.; CARVALHO, P.; CORRÊA, V. A; CARDOSO, A. V. **Golden mussel: detection of a dangerous invader**. 2016. Available at: < https://www.researchgate.net/publication/304930013_Mexilhao-Dourado_no_Brasil_Deteccao_de_um_perigoso_invasor>. Accessed on: 06 July 2016.

SILVA, M. R. **Evaluation of the use of vinasse and *Limnoperna* fortunei (Dunker, 1857) residue in sugarcane soil: possibility of replacing nitrogen, phosphorus, potassium and micronutrients.** Araçatuba, São Paulo, 2010. Monograph (Degree in Biofuels Technology) - Faculty of Technology (Fatec).

SINDMAR - National Union of Merchant Navy Officers. **Role of seafarers highlighted on World Day.** Available at: < http://www.sindmar.org.br/noticias/pagina/217>. Accessed on: 23 July 2016.

PAULISTA ENVIRONMENTAL SYSTEM. **Invasive alien species are the second biggest cause of biodiversity loss.** 2009. Available at:< http://www.ambiente.sp.gov.br/2009/05/25/especies-exoticas-invasoras-are-the-second-largest-cause-of-biodiversity-loss/>. Accessed on: 11 July 2016.

QUITINO, C. A. A. A **history of environmental education in Brazil and worldwide.** 2013. Available at: <http://www3.unifai.edu.br/pesquisa/publica%C3%A7%C3%B5es/artigos-cient%C3%ADficos/professores/bacharelados/um-hist%C3%B3rico-sobre-educa%C3%A7%C3%A3o>. Accessed on: 10 July 2016.

VALÉRY, L.; FRITZ, H.; LEFEUVRE, J.C. SIMBERLOFF, D. 2008. **In search of a real definition of the biological invasion phenomenon itself. Biological Invasions,** 10: 1345-1351.

VERMULM JÚNIOR, H.; GIAMAS, M. T. D. **Occurrence of the golden mussel *Limnoperna fortunei* (Dunker, 1857) (Mollusca; Bivalvia; Mytilidae), in the digestive tract of the "Armai" *Pterodoras granulosus* (Valenciennes, 1821) (Siluriformes; Doradidae), from the Paraná River.** São Paulo, Brazil. São Paulo: B. Inst. Pesca, 34(1): 175 - 179.

2008.

VILLAR, L. S. **Optimisation of methodology for the extraction and determination of metallothioneins by polyacrylamide gel electrophoresis from *Crassostrea Rhizophorae* (Guilding, 1828) oysters for application in environmental studies.** 58f. Dissertation (Master's in Chemistry) Fluminense Federal University. Niterói, Rio de Janeiro, 2006.

VON RÚCKERT, G.; CAMPOS, M. C. S.; ROLLA, M. E. Feeding *of Limnoperna fortunei* (Dunker, 1857): filtration rates with emphasis on the use of *cyanobacteria.* **Acta Scientiarum Biological Sciences,** Maringá PR, v.26, n.4, p.421-429, 2004.

ZANELLA, O.; MARENDA, L. D. Occurrence of *Limnoperna fortunei* in the Itaipu Hydroelectric Power Station. In: **5th Latin American Congress of Galactology.** Instituto Butantã/ Instituto de Biociências - USP. São Paulo, Brazil, 2002.

More Books!

yes

I want morebooks!

Buy your books fast and straightforward online - at one of world's fastest growing online book stores! Environmentally sound due to Print-on-Demand technologies.

Buy your books online at
www.morebooks.shop

Kaufen Sie Ihre Bücher schnell und unkompliziert online – auf einer der am schnellsten wachsenden Buchhandelsplattformen weltweit! Dank Print-On-Demand umwelt- und ressourcenschonend produzi ert.

Bücher schneller online kaufen
www.morebooks.shop

info@omniscriptum.com
www.omniscriptum.com

OMNIScriptum

FSC
www.fsc.org
MIX
Papier | Fördert
gute Waldnutzung
FSC® C083411

Kolibri 360 GmbH
Ferdinand-Jühlke-Straße 7
99095 Erfurt, Deutschland
produktsicherheit@kolibri360.de

Druck:
CPI Druckdienstleistungen GmbH
im Auftrag der
Kolibri 360 GmbH
Ferdinand-Jühlke-Str. 7
99095 Erfurt